Farm Animals

Goats

Heather C. Hudak

Published by Weigl Publishers Inc.
350 5th Avenue, Suite 3304, PMB 6G
New York, NY 10118-0069
Website: www.weigl.com

Library of Congress Cataloging-in-Publication Data

Hudak, Heather C., 1975-
Goats / Heather C. Hudak.
p. cm. -- (Farm animals)
Includes index.
ISBN 1-59036-424-4 (hard cover : alk. paper) -- ISBN 1-59036-431-7 (soft cover : alk. paper)
1. Goats--Juvenile literature. I. Title.
SF383.35.H83 2006
636.3'9--dc22

2005034681

Printed in the United States of America
1 2 3 4 5 6 7 8 9 0 10 09 08 07 06

Editor Frances Purslow
Design and Layout Terry Paulhus

Cover: Goats are very hardy animals. They can live in many different environments.

Contents

Meet the Goat

Goats are small farm animals. They can have short or long hair, which can also be curly, silky, or coarse. Goats have **wattles** and short tails. They may have straight or curved noses. Their ears may be pointy or droopy, small or large. Goats come in many colors, shapes, and sizes.

Goats are mammals. Like other mammals, goats have hair on their body. Mother goats feed their young with milk from their bodies.

Goats are active animals. They jump, climb, run, and crawl. Some goats can jump as high as 6 feet (1.8 meters) in the air.

Goats are curious and smart. Goats explore new things by sniffing and nibbling. They can also be trained.

Some people confuse goats and sheep. Goats have beards. Their horns curve backwards. They also have straighter hair and are thinner than sheep.

Goats with horns
are sometimes
called "buttheads."

All about Goats

Goats are social animals. They live in groups called herds. There can be as many as 20 goats in a herd. Goats talk to other goats by bleating, or making a "baah" sound. Goats will bleat when stressed, hungry, or calling for attention from their keepers. Mother goats bleat when calling to their young.

There are eight **species** of goat. Some of the most common farm **breeds** include the Angora, Kashmir, and French Alpine.

There are more than 210 different breeds of goats.

Breeds of Goats

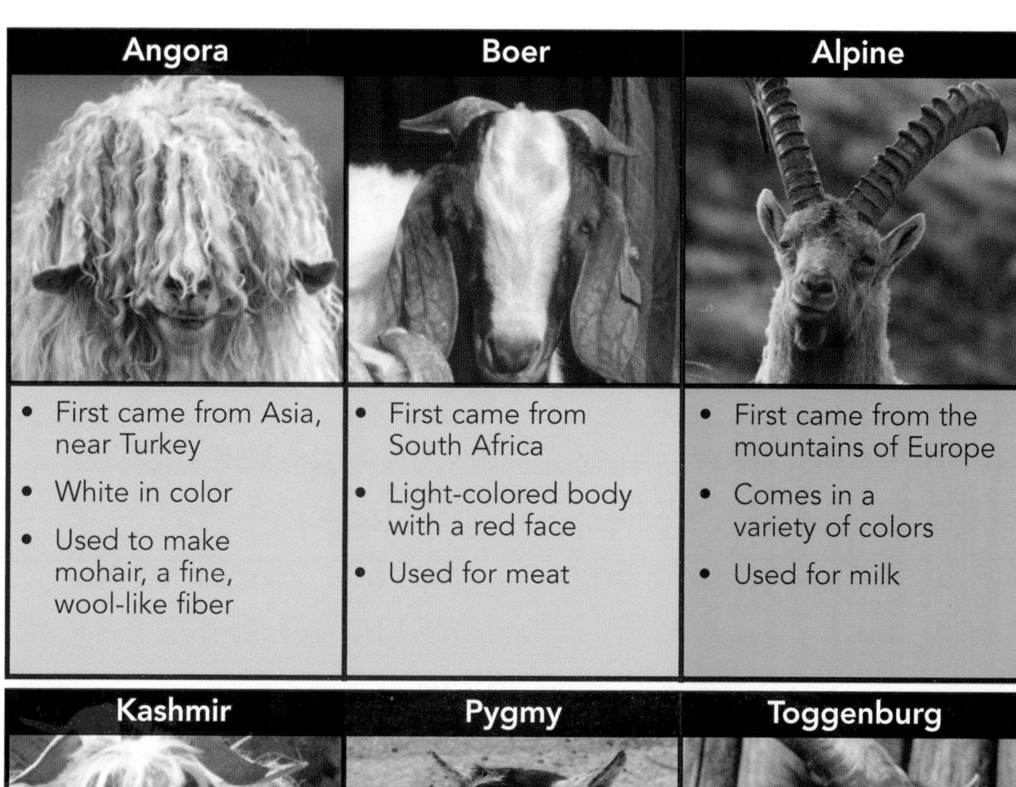

Angora	Boer	Alpine
• First came from Asia, near Turkey • White in color • Used to make mohair, a fine, wool-like fiber	• First came from South Africa • Light-colored body with a red face • Used for meat	• First came from the mountains of Europe • Comes in a variety of colors • Used for milk

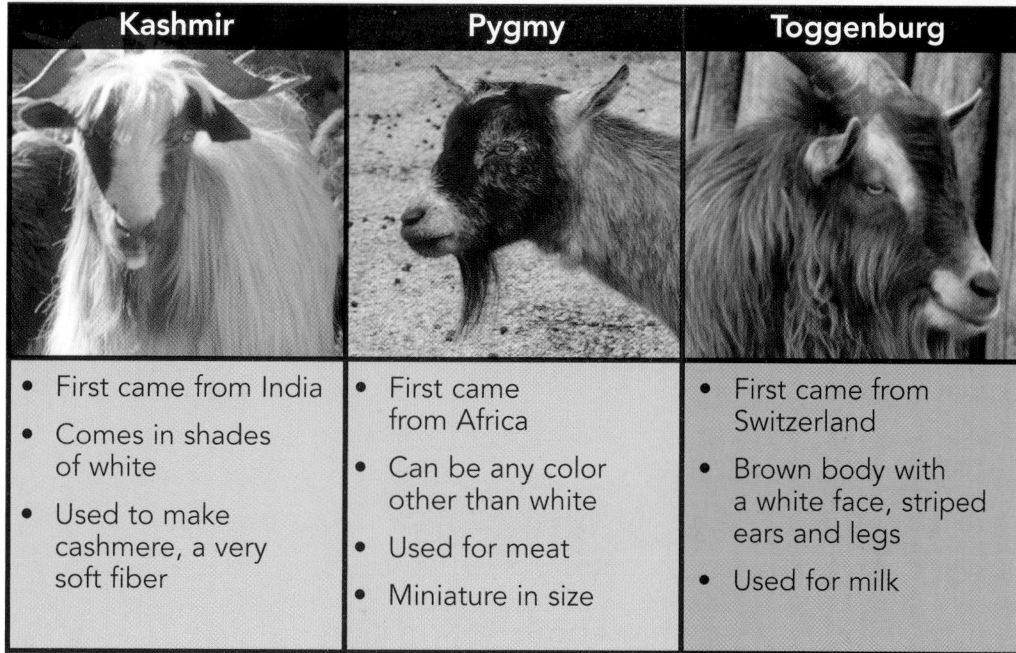

Kashmir	Pygmy	Toggenburg
• First came from India • Comes in shades of white • Used to make cashmere, a very soft fiber	• First came from Africa • Can be any color other than white • Used for meat • Miniature in size	• First came from Switzerland • Brown body with a white face, striped ears and legs • Used for milk

Goat History

Goats have existed for thousands of years. Goat **fossils** found in Asia are more than 9,000 years old. Goats were among the first animals tamed by people. The first tamed goats came from Iran.

In the 1500s, Spanish explorers brought the first goats to North America. Shortly after, English settlers also brought goats to the New World. In the 1900s, goats were brought to the United States from Europe for use as **dairy** animals.

Before AD 500, goatskin was used to make water and wine containers. It was also used as paper. People ate goat meat and drank goat milk. Today, wool from goats, such as mohair or cashmere, is used to make clothing and other cloth items.

Today, there are about 450 million goats on Earth. Between 2 and 4 million goats live in the United States.

The ibex is a wild goat that lives in the mountains of Europe, Asia, and Africa.

Goat Shelter

Farm goats need a clean, dry shelter. They need a place that gives them shade from the Sun and keeps them warm on cold days. The shelter should protect goats from wind, rain, and snow.

If goats are kept indoors, their stalls should be at least 12 square feet (4 square meters). If their stalls are too small, goats will become upset. Goats should be able to see other animals, too. **Slatted** walls help them do this.

Goats need daily exercise. They should be allowed outdoors every day to **graze** in large fields called pastures. Goats are very curious. Pastures should have fences to keep goats from wandering away.

Cedar shingles can be used to make a goat shelter.

Goats are very smart. They can learn to open the latches on pens and gates.

Goat Features

Goats are **sturdy** animals. Their bodies are **adapted** to living on hills, mountains, and farms. A goat's woolly undercoat keeps the goat warm in cold weather. Goats have a light build. This helps them move quickly and effortlessly. Their **surefootedness** helps them leap easily from rock to rock. Other parts of a goat's body also have special features and uses.

STOMACH
Goats are ruminants. Ruminants have stomachs with many parts. Goats have a four-part stomach to help them break down their food.

FEET
Goats have hoofed feet. Hooves are horn-like shells that protect the soft parts of a goat's feet. Each hoof has two toes with a rough pad on the bottom.

HORNS

Most goats have horns. Goat horns are hollow. They never stop growing. Male goats have larger horns than females.

EYES

Goats have square **pupils**. This helps them see well at night.

TEETH

Goats have 24 molars in the back of their mouth.

BEARD

Both male and female goats can grow beards. Beards on male goats grow longer than those on females.

What Do Goats Eat?

Goats are herbivores. Herbivores eat plants. Goats eat grass, herbs, shrubs, weeds, hay, and tree leaves. When eating, goats swallow their food without much chewing. Then, they spit it up into their mouths. This is called cud. Goats chew the cud again before swallowing a second time.

Farmers often feed goats hay. Goats get extra energy from grains, such as oats, barley, and corn. Crushed peas and soya beans add **nutrients** to their diet. Goats also need fresh, clean water to drink.

Goats may die if they eat certain plants. These plants include nightshade, yew, laurel, bracken, and dead fruit-tree leaves. Mold and corn stalks can also make goats ill.

Goats are selective eaters. They will not eat food that has fallen on the ground or that has been touched by another animal.

Goats have very good balance. Sometimes, they can even climb trees.

Goat Life Cycle

Mother goats are called does or nannies. Adult male goats are called bucks or billy goats. Baby goats are called kids.

Nannies carry their babies in their belly for about five months. When they give birth, it is called kidding. Most nannies have one or two babies at a time.

Newborn

At birth, kids weigh 2 to 4 pounds (0.9 to 1.8 kilograms).

Young goats drink their mother's milk. It has many healthy nutrients to help kids grow big and strong.

6 Months to 2 Years Old

Kids are weaned, or stop drinking their mother's milk, when they are about 3 months old. Goats can begin to have babies at about 18 months old. At this age, they weigh 55 to 65 pounds (25 to 29 kg).

Kidding lasts between 1 to 2 hours. Twins are born within 5 to 30 minutes of each other.

Nannies often give birth to kids in the springtime. Kids grow quickly. With special care, goats live for about 8 to 15 years.

Adult

Goats grow until they are 30 months old. They can grow up to 42 inches (1.1 m) tall at the shoulder. Most full-grown farm goats weigh between 100 and 200 pounds (45 and 90 kg).

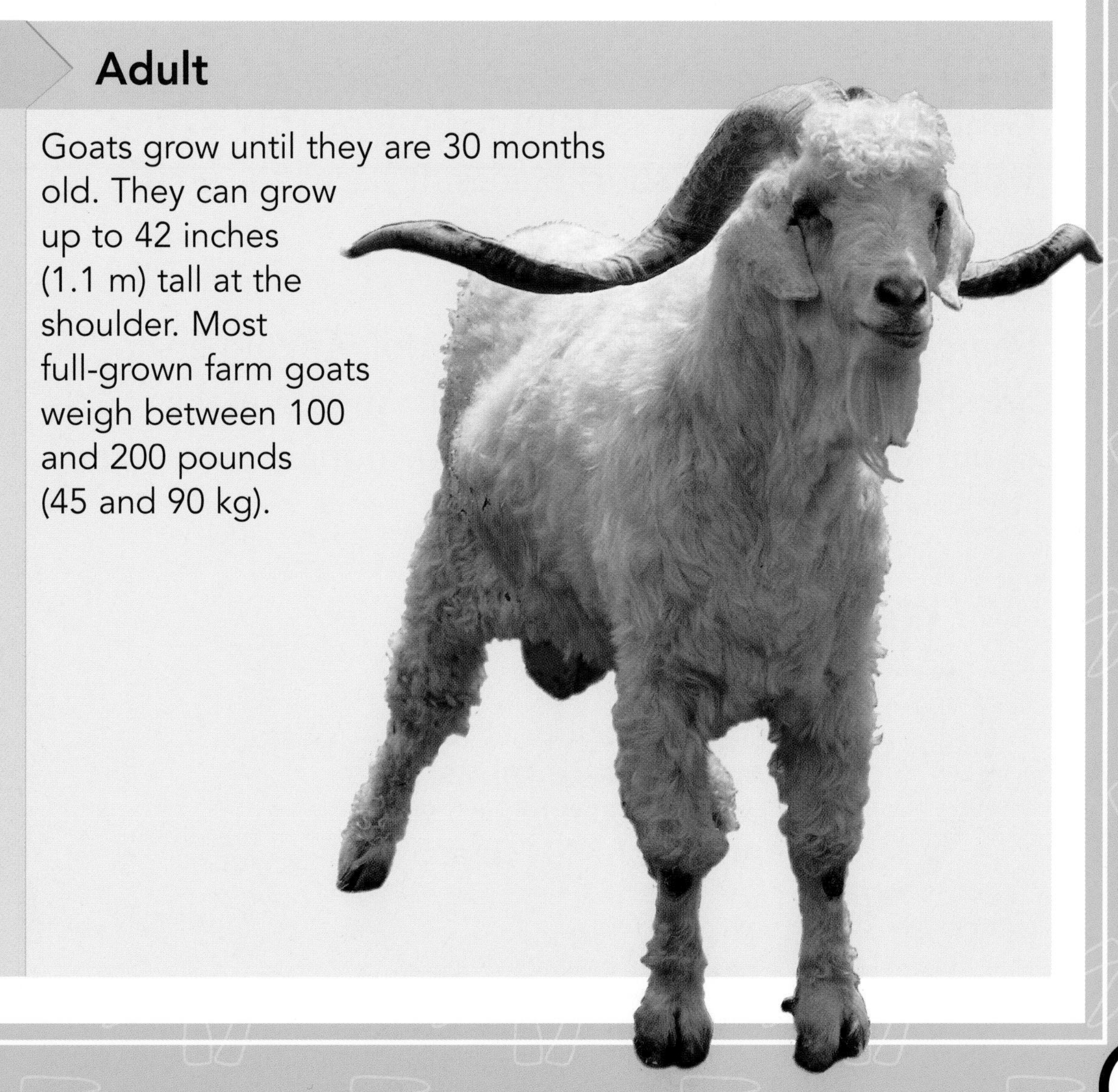

Caring for Goats

Goats need special care. They must visit the veterinarian, or animal doctor, at least once a year. At the vet, goats are given **vaccinations**. They should also be **dewormed**. At home, farmers should regularly check their goats for lice. Lice are insects that can harm a goat's skin and hair. They can also cause goats to develop **anemia**.

Goats need their hooves trimmed often. This is done with sharp shears or short blades. It keeps the goat's feet and legs healthy.

Some goats are sheared for their wool. Angora goats are sheared twice a year. Each adult Angora makes about 8 to 16 pounds (4 to 7 kg) of mohair each year.

Useful Websites

To learn more about goat farms, visit: **www.mda.state.mi.us/kids.** Click on "County Fair," then "Animals," and "Goats."

Goats make good pets because they are naturally curious and friendly.

Myths and Legends

For thousands of years, many cultures have told tales about goats. The Chinese have a special calendar that links each year with a certain animal. People born in the year of the goat are thought to be calm, patient, and creative.

For thousands of years, some people have believed that a particular group of stars in the sky is shaped like a goat.

This goat is called "Capricorn." Those born between December 22 and January 19 are said to be born under the sign of Capricorn.

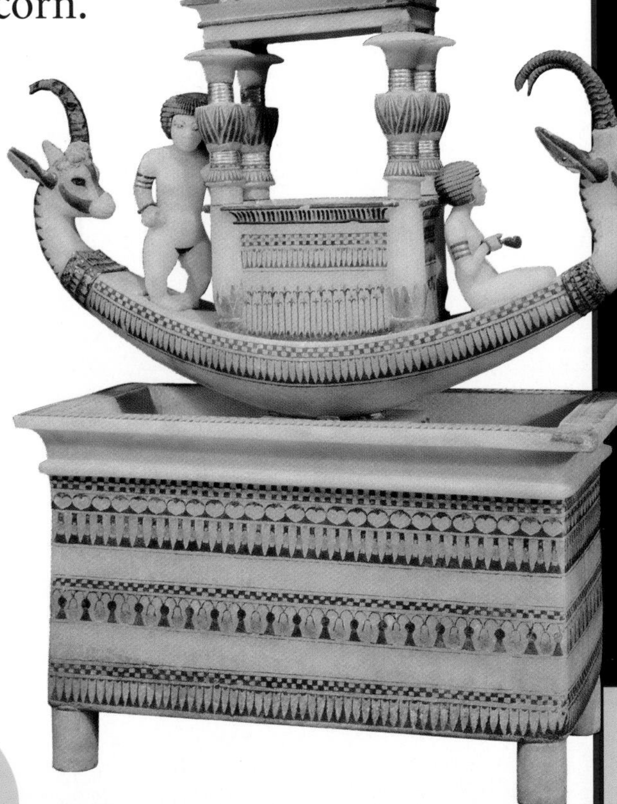

Ancient objects honoring wild goats have been found in Egypt.

Rolling Chariot

Thor was the **Norse** god of thunder. He was the protector of gods and humans against the forces of evil.

The Norse people believed that during a thunderstorm, the sound of thunder was caused by Thor riding through the heavens on his **chariot**.

Two magical goats pulled Thor's chariot. One was named *Tanngrisni*, meaning "gap tooth." The other was named *Tanngnost*, meaning "tooth grinder." Earth burned and the mountains cracked as the two goats ran across the sky.

Frequently Asked Questions

How can I protect my goats from predators?

Answer: Having guard dogs is a good way to keep goats safe. Great Pyrenees and Anatolian shepherds are good guard dogs. Llamas and donkeys can also be used to help protect your herd.

How much milk do dairy goats produce?

Answer: Many dairy goats produce about 1 gallon (4 liters) of milk each day. Some goats can produce about 600 gallons (2,270 L) of milk in one year.

How many goats can a farmer own?

Answer: Most farms can have six to ten goats per acre (0.4 hectare) of land. If there is a large amount of grazing land in the area, a farm can have more goats.

Puzzler

See if you can answer these questions about goats.

1. What are baby goats called?
2. Where did goats first come from?
3. What can goats be used for?
4. What do goats eat?
5. Where do tamed goats live?

Answers: 1. Kids 2. Asia 3. Meat, milk, and fiber 4. Grass, herbs, shrubs, weeds, hay, and tree leaves 5. On farms and pastures

Find Out More

There are many more interesting facts to learn about goats. If you would like to learn more, take a look at these books.

Wolfman, Judy. *Life on a Goat Farm.* Carolrhoda Books, 2002.

Damerow, Gail. *Barnyard in your Backyard.* Storey Books, 2002.

Words to Know

adapted: adjusted to the natural environment
anemia: a blood disorder
breeds: groups of animals that have common features
chariot: a cart with two wheels that was used in ancient times
dairy: used for millk
dewormed: cured an animal of a disease caused by worms
fossils: preserved remains of living things
graze: to eat grass in a field
Norse: ancient Scandinavians
nutrients: parts of food that nourish living things
pupils: parts of the eye through which light passes
slatted: made with strips of wood with spaces between
species: living things that can mate with one another but not with those of other groups
sturdy: strong
surefootedness: unlikely to stumble or fall
vaccinations: medicines that prevent disease
wattles: hair-covered folds of skin located under the chin

Index